Swift Summers

My life with the Common Swift

Mark Walker

© 2016. Mark Walker. All rights reserved.
ISBN 978-1-329-96309-2

Author contact: mark_david_walker@yahoo.co.uk

Common Swifts are beautiful birds. This beauty comes from their perfectly aerodynamic profile. We instinctively admire such perfection. But few people get the chance to examine these birds closely because of their aerial lifestyle. I was lucky enough to have the opportunity to work with these special birds at close range. Between 2006 and 2011 I had access to a nesting colony where I could handle them daily. Over the summer months I would spend most daylight hours watching, observing and working around these amazing birds gaining a unique perspective into their lives.

This booklet is a collection of personal observations about the swifts at the most vital part of swift life; the summer breeding season. I have tried to convey information about swifts in a more personal and interesting style than is the case in dry scientific texts. Through doing so I hope to introduce you to these amazing birds and pass on my some of my love for them.

Swift Summers

Contents	Page
Introduction	7
Chapter 2: Super Swifts	23
Chapter 3: First Contact	43
Chapter 4: House Building	63
Chapter 5: Young Nestling	81
Chapter 6: Life in the Nest	97
Chapter 7: Teenager Troubles	111
Chapter 8: Fledging	127

Chapter 1

Introduction

On warm summer days high above the city streets, way above bustling people, and even higher than the tallest office blocks, Common Swifts fly adeptly in a world of their own. Their long arced wings make them instantly recognisable. Busy commuters and shoppers can be forgiven for not noticing such seemingly unobtrusive birds that seem so remote to what is happening in their own lives.

Only when swifts fly closer to the ground to reach low nesting sites do they become more obvious, even to the most unobservant of city dwellers. Their manoeuvrability and speed in flight is amazing and somewhat reminiscent of aircraft formation flying. Common Swifts swoop and loop in ways hardly imaginable to us ground bound humans. Their harsh calls puncture the silence of summer days and seem to cockily mock our aerial inability.

The chance to watch Common Swifts at close range is a relatively rare event. The Common Swift is migratory and visits us in our northern climes for on average only 100 days each year, the rest of the time it is resident in sub-Saharan Africa. For much of the period it is with us it flies way above our heads out of sight. Nesting sites are commonly found in high buildings. Swifts are particularly vocal at the beginning of the mating season when courting, but once incubation commences they diligently carry out their egg warming duties and become much more inconspicuous.

The Common Swift is known scientifically as *Apus apus*. Despite being common and numerous, it has been relatively little studied. There are very good reasons for this. One is that the biological research world is rather strangely skewed. You would expect there to be most biologists where biological diversity is at its highest, which is in the tropics and equatorial areas. Counter intuitively most biologists have their academic positions in North America and Europe where biological diversity is much lower. Although the Common Swift spends two thirds of its life in Africa it has hardly ever been studied there. Practically all studies are undertaken in Europe, by European scientists. As the swift visits us for such a short time each year, this limits the research that could be done on it. Although admittedly the most interesting aspects of swift biology, it's breeding, takes place in Europe.

How to weigh a swift!

However perhaps the most important reason why there is a paucity of swift research is that the swift is simply not easy for the ground bound biologist to get at. The problem is that it has become such a specialized flyer that it nearly never lands. It can even sleep on the wing. Unless a biologist can actually get a bird in the hand there is little he can actually study. Studying birds of whatever species is problematic because of their aerial skills.

A wandering nestling.

This means that the only realistic chance a biologist has of studying the swift is when it does actually have to visit the terrestrial world. In other words when it breeds. Actually being on the ground must be a very strange experience to a swift which spends nearly its entire time in the air. It must be a very eerie uncomfortable experience for the birds to find themselves limited to two dimensions instead of three. Maybe it is something akin to us going snorkelling.

On the ground as well as unpleasant biologists trying to catch them, there are a whole host of nasty predators who would love to meet up with swifts and invite them to dinner. Therefore it pays swifts to lose the cockiness they exhibit in the air and to be more cautious and timid on the ground. They search out nest sites that allow easy entry to themselves but minimum access to predators. This means that nests are invariably built in high up places and are found mostly amongst small nooks and crannies. But what is inaccessible to a predator is also inaccessible to a human researcher. So biologists are in the frustrating situation that they can often see where swifts nest but have no physical way of getting to them.

Only a handful of locations allowing hands on study have ever been discovered. Maybe the most ingenious was the nesting colony developed by the Swiss Ornithologist Weitnauer in the 1940's. On finding Common Swifts nesting near to his house he renovated his roof to provide more nest places for them. He designed these nest boxes in such a way that he could observe the birds on the nest. He spent many hours in his attic observing the antics of the swifts and was able to provide the first detailed account of their nesting behaviour.

Another well-known example of a place where Common Swifts have been studied is in Oxford, England. Swifts began squatting in the university museum roof soon after it was built. They enter through ventilation holes. In the 1950's the ornithologist David Lack had the roof and the nest sites adapted to allow him and his students to study the swifts as they nested without the birds being disturbed. This allowed Lack to collect many interesting and useful observations about the swifts. His work is recounted in the beautifully written and eloquent book 'Swifts in a Tower'. This book has remained a first place of reference for all people interested in Common Swifts.

Although other researchers have studied the Common Swift, none have had such easily accessible research sites as these, nor have been able to study the swift in as much detail as has been possible at these two locations. Most other studies of the swift have been done through mist netting swifts close to their nesting roosts and have not been done over the long term.

So when I heard that there was an easily accessible nesting colony close to where I lived in Germany I jumped at the chance to find out more. Local bird watchers had noticed swifts flying into and out of a concrete road bridge spanning a local reservoir. Out of interest they had contacted the road authority who had responsibility for the bridge and who had allowed them access to the inside.

On visiting the bridge it was apparent to them that the swifts had found an ideal new nesting site. There were several nests lying safely inside enclosed concrete walkway running underneath the bridge. They contacted a biologist at the local university in Marburg. He immediately recognised the potential of the site, but because of work commitments was not able to start a study at the bridge himself. Instead I was asked if I wanted to use the site. I leapt at the chance.

Motorway bridges in Germany are given names, and my bridge was named the Ronnewinkel Brücke (Brücke is German for Bridge). It carries a Bundesstrasse, a German main road, across a large man made reservoir called the Bigge or Biggesee. The bridge is roughly 500 metres long from one end to another.

I first visited the bridge on a bright autumn day in 2006. It was a busy location with constant noise from the traffic. The bridge has two internal wakways, one for each carridgeway, running underneath the road along its entire length.

The exterior of the bridge.

The area around the bridge was particularly picturesque. This area of Germany is known as the Sauerland. It is a heavily forested, hilly region, popular with weekend visitors from the Netherlands or the nearby industrial Ruhr towns looking for a peaceful daytrip in the country. Although the Bigge reservoir was built to supply water to the thirsty millions in the Ruhr valley it quickly became a magnet for tourists. Nothing is as relaxing as a day spent on the grassy banks of the lake swimming or sunbathing. The reservoir is deceptively large, because rather than being circular in shape, it instead winds its way through the valley in a serpentine manner.

At the end of the bridge was a narrow path leading beneath the roadway for private access. I walked down the steps and approached one of the metal entry grills which were situated under the carridgeway. I unlocked the padlock with the key I had been given and climbed in.

It was very dark in the bridge, however after a few minutes my eyes adjusted somewhat to the dim conditions. Although I could not see perfectly, I could at least make out some of the walls. I had a head torch for light. This was comfortable to wear. In fact it was a little too comfortable. Many months later I forgot I had it on my head and wandered about outside the bridge for hours wondering why all the people I passed were looking at me funnily.

The only natural light in the bridge came from the holes which had been made on the floor. These were probably made for drainage in case water should run into the bridge. The floor was simply a thick sheet of concrete, and the holes had been bored through to the outside. The holes were about 10 cm wide and varied from 7 to 60 cm in depth. When the sun shone a shimmering shadow from the water below was cast inside. By looking through the holes the water below could be clearly seen. It is through these holes which the swifts scrambled to reach the nests. The overall impression I got was one of darkness, dinginess and dustiness.

An inside chamber.In winter these were very cold but they became very hot in summer when they trapped the heat.

The walkway ran underneath the entire bridge. There was one for each carridgeway. Each was divided up into a number of separate chambers. Each chamber was 40 to 48 metres long and was seperated from the adjoining chambers by a thick concrete wall. Altogether each side of the bridge was made up of eight chambers, meaning that there are 16 chambers in total. Within each dividing wall was a small roughly one metre wide hole through which you could climb to reach the next chamber. Chambers were not exactly rectangle shaped because the width changed along the course of the chamber to follow the curvature of the bridge.

The road surface was probably only a few centimetres above the ceiling of the chambers. The noise from traffic was almost constant. Heavy lorries in particular were the noisiest, and made the floor of the chamber shake and wobble so that you could feel the movement through your whole body. The floors of the chambers were not totally level. Much of the floor was covered with a thick layer of dust, and this appeared to originate from the concrete itself rather than as dirt which the swifts bring into the bridge. After working in the bridge I would be covered with a layer of pale white dust.

The Bigge Reservoir.

As I walked between the chambers I noticed that besides many of the holes there were nests. However many holes did not have any. Typically there were one to seven nests per chamber. Nests appeared to be made of grass that had been matted and woven together. Some appeared old and in poor repair, but others were obviously in good condition and more recent.

What attracted the swifts to this location? This was an easy question to answer. You only had to go down to the waters edge of the reservoir in the evening to see cloud like masses of insects coming from the water. One could almost call it aerial plankton. By nesting in the bridge the swifts had a larder of easily accessible food on their doorsteps. However the swifts don't have this veritable feast all to themselves. There are many competitors they have to share this valuable resource with. The Biggesee is also home to substantial numbers of swallows and house martins.

After my first visit to the bridge the possibilities that the site offered were apparent. I could easily get into and out of the bridge to study the swifts. No climbing up to high roofs or onto precarious high precipices was necessary. The swift nests were situated on the ground where I could easily get to them. Inside the bridge there was space for setting up equipment. Although the darkness inside the bridge hindered easy working, it meant the swifts could not see me, and this meant I could get unbelievably close without disturbing them. The swifts would literally fly into my hands. Over the following summers I would learn to love the swift and experience its life at first hand.

Chapter 2
Super Swifts

Most of the world's swift species might at first sight appear rather drab, dull and uninteresting. Many are a positively underwhelming greyish-black in colour and rather small in size. Our common or European Swift is conspicuously devoid of decoration. However despite appearing rather boring, appearances can be deceptive.

Swifts are in fact some of the most interesting birds within the avian class, possessing a range of abilities and talents of almost super-natural proportions. They are in fact swift super-heroes. But before we look at some of the swifts supernatural abilities, it is important that we have a look at what swifts there are and how they are related to each other.

The swift family tree

There are two families of swifts; the more showy Tree Swifts, (Family Hemiprocnidae) which have head crests, and the True Swifts (Family Apodidae). Traditionally these two families have been placed in the Order Apodiformes along with the Hummingbird family (Trochilidae) and sometimes also the Nightjars (Caprimulgiformes) as these were thought to be most closely related to the Hummingbirds. The Apodiforme Order was named by the German ornithologist Ernst Hartert in 1897.

Some taxonomists now place both Hummingbird and Swift families within a single super order known as the Trochiliformes. Although swifts might look like a number of other species of birds, such as Swallows or House Martins, they are in fact not closely related to these species. Swallows and House Martins instead belong to the large Passerine Order of birds.

The Apodidae is divided into two subfamilies, the primitive American Swifts the Cypseloidinae and larger more advanced Apodidae. There are 13 species in the Cypseloidinae. In comparison to the Cypseloidinae the

Apodinae are much more cosmopolitan and more advanced. The Apodinae subfamily is divided into three tribes. The Typical Swifts (Apodini), the Swiftlets (Collocalini) and the Spinetails (Chaeturini).

The Swiftlets are found mostly in south-east Asia, although some species can be found in Africa and the Indian subcontinent. The Spinetails are found across the Americas. The Typical Swifts have on the other hand, a worldwide distribution, with species being found on every continent apart from Antarctica. The Typical Swifts are sometimes considered as being the most advanced of the swifts, with the Swiftlets and Spinetails being a kind of intermediate between them and the primitve Cypseloidinae.

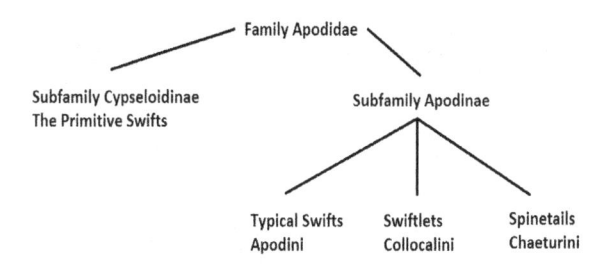

```
                        Family Apodidae
             /                              \
Subfamily Cypseloidinae              Subfamily Apodinae
The Primitive Swifts                /        |         \
                          Typical Swifts  Swiftlets  Spinetails
                          Apodini         Collocalini  Chaeturini
```

Deciphering species relationships within the Apodiformes has been difficult due to the extent of convergent evolution which has taken place. Species are often similar in appearance. The demands of an aerial lifestyle place constraints on possible body designs and this has resulted in many species looking the same.

The swift fighter jet
Maybe the most obvious of the swifts supernatural abilities is its rather amazingly fast and manoeuvrable flight. The swift makes full use of its perfectly adapted form and has become the ultimate aerial specialist amongst the bird kingdom. Swifts are excellent fliers able to fly at speeds of up to 200 kilometres per hour. What adaptations do they possess which allow this? In what ways are the wings and bodies adapted to allow such excellent flight?

A human engineer tasked with designing a new fighter jet would do well to study the swift for inspiration. Both fighter jet and swift are optimally designed for high speed predatory flight.

The typical swift profile makes it superbly aerodynamic.

In form the swift is perfectly aerodynamic being torpedo shaped in profile. The head is short and rounded. There is no neck to talk of; the head being rather stiffly joined to the body. Although this might make sense aerodynamically by reducing drag during flight, it means the swift is unable to turn or twist its head. When the swift wants to look at something it has to move its entire body so that it is directly facing it.

The beak is short and black. The eyes are positioned on the side of the head, but the swift still has excellent forwards vision. The feathers directly in front of the eyes are depressed slightly allowing good forward visibility.

The front part of the body, around the chest, has the greatest circumference. The body gradually and smoothly tapers away until the end of the tail. There are no protruding extremities that would produce drag and hinder the swift in fast flight. The feet are kept unobtrusively tucked away. The wings, as already mentioned, are swept backwards in a long slender scimitar like arc.

If you see swifts on the ground one of the first things you will notice are the rather ungainly long wings. These are very long compared to those of other birds. Bird wings are conventionally divided into 'hand' and 'arm' sections, the hand section being composed of the carpometacarpus and digits, and the arm being made up of the humerus and ulna/radius. The dimensions and sizes of these bones vary between different bird species, reflecting the different lifestyles they have. Common swifts have extremely long hand wings, which are 75% the length of the total wing. The primaries are longer than is seen in other types of birds, especially in proportion to the secondary's, and this is because the carpometacarpus in swifts is extremely long.

Flight feathers are known as remiges. The primary remiges, or primaries, are found in the hand area of the wing and number between nine and eleven. These are strong and heavily asymmetrical. The one at the leading edge, number XI, is only two centimetres in length, stiff and without vanes. It acts to support the base of the longest primary, number X.

The secondary remiges are found in the arm part of the wing and the number varies greatly between bird species. There are seven secondary feathers in the swift. The feathers on the front leading edge of the wing are known as the alula. In the swift this is about an eighth of the length of the total wing, and is made up of two or three feathers.

Although many species of seabirds, such as the albatross, spend an equally substantial amount of time on the wing as the swift, these have an enlarged arm wing rather than hand wing for the generation of lift.

Swifts have wings with high aspect ratio values. The aspect ratio is a measure of the longness and narrowness of wings and can be found by dividing the mean wingspan by the wing area. Long wings with a high aspect ratio are good for gliding flight and produce less drag, but shorter wings are easy to flap and allow more manoeuvrability.

Birds which fly long distance or are aerial specialists and which fly in open habitats, such as the swifts, have wings with high aspect ratios. Birds which on the other hand fly in 'cluttered' environments, for example through bushes or shrubby undergrowth, such as the wren or the sparrowhawk, have wings which are shorter and have lower aspect ratios.

The wing loading is the ratio of weight to the area of the wings. Species with low wing loading, in other words large wings for their weight, fly slowly and flap the wings less rapidly. Species with high wing loadings have small wings for their weight, and this means they have to flap the wings faster and therefore fly at a greater speed. Swifts have high wing loading resulting in rapid flight.

This photo by Klaus Rogge shows the wing feathers well.

Swift feet

For some reason swift biologists have been inordinately interested in the swift feet, using them as a characteristic to taxonomically divide swift species. Swifts have small feet. The family name of the European Swifts, Apodidae, actually means 'without feet' in Greek. This is actually a good name as swifts have only very small feet and these are not apparent while the swift is in flight. The feet are retracted and kept close to the body during flight reducing drag to a minimum. It is easy to imagine the disadvantage in aerodynamic terms of having long dangling legs.

Many species of swift are unable to perch with their feet because of their shortness. The swift, for example, must 'belly flop' around when on the ground because its legs are totally unsuitable for supporting the weight of the body or for forward propulsion. This does not mean however that the legs and feet are totally weak. The feet end in sharp stabbing claws and can grip with surprising firmness.

Usually birds have four toes. The fifth toe has regressed to at most a foot spur or has disappeared completely. The first toe points backwards and is known as the toe bone. The other three point forwards and are numbered from two to four, counting from inside the foot outwards. A variety of foot 'designs' occur in the animal kingdom and have often been used to classify birds.

Members of the Apodidae show a variety of foot designs. The Cypseloidinae are anisodactyl, which means that they show the 'normal' pattern of foot shape, with digit one being pointed backwards and digits two, three and four pointing forwards. The Apodini are heterodactyl, with digits one and two pointing forwards and digits three and four which point backwards. However many Apodini have extremely flexible feet and are able to move their toes into a variety of positions to allow them to grasp onto a variety of surfaces. This has led some to classify some species, such as the Chimney Swift as being pamprodactyl, with having all their toes pointing forwards. Some exhibits in museums are still shown in this way, although most biologists now agree that this does not reflect the true nature of the swift foot. The Chaeturini and Collocaliini are anisodactly like the Cyseloidinae.

Some superhuman swift feats
Swifts are the superheroes of the avian world. They possess a range of almost supernatural abilities that any aspiring comic hero would be proud of.

The amazing climbing spinetail swifts: Swifts are able to walk up walls. The tails are short, but stiff and forked. The stiffness of the tails is of advantage when birds are hanging on cliff sides or house walls, as they can be used as a prop. There are normally ten tail feathers, although there can be more.

Some of the spinetail swifts have evolved even better tails with extended rachi, the internal shafts, which protrude from the tips of the tail feathers like running spikes and thus provide grip when climbing. This is seen in species belonging to the Chaeturni tribe.

Never ending flights: The swift has taken its aerial propensity to almost absurd proportions. Many species hardly ever land. They can do everything in the air. As an aerial insectivore, feeding is, of course, done in the air. Lakes can be skimmed in order to drink.

Maybe the most spectacular feat of swifts is the apparent ability to fly on auto-pilot while they sleep. The swift apparently lets one half of its brain take charge of the flying and all the other bodily processes, while the other half of the brain is switched off and does some sleeping. Later the roles are exchanged and the other half of the brain gets some sleep.

Of course the problem with flying while sleeping is that if you fly in a straight line, when you wake up you are in a different place from where you started from. The swift solves this by flying in a circle. When there is wind they fly into it. The swift is still displaced somewhat, but being such an excellent flyer, it does not take it long to fly back to where it wants to be.

Another example of its aerial lifestyle is shown by reports that swifts mate in the air. An extreme example of the mile high club! Although not confirmed, many bird watchers have reported seeing two birds coming together in mid-air. Various anatomical parts make contact which would seem to indicate that some kind of copulation is taking place.

However, whether this is actually the case is difficult to determine because of the distance from the observers these events are normally observed. Plus male and female swifts are identical and the sexes can't be identified. However, because of the frequency of which this is reported to occur, with swift watchers sometimes seeing it more than once per season, it really does seem as though it does occur. Despite this most mating does seem to occur at the nest on solid ground.

Swifts often nest in high inaccessible locations such as in chimney stacks or in high roofing guttering. This means when the time comes for nestlings to leave the nest, they have to be able to fly perfectly on the first attempt, without practise, and without any problems. Failure is not an option. That nestlings manage to make it into the air is in itself a pretty amazing feat. Mostly the nestlings seem to master flight perfectly by instinct on this first attempt being able to swoop and dive without effort or conscious thought.

But that is not all. This maiden flight is not just a few seconds of scrabbling before the next attempt is tried, these first flights can last considerable lengths of time. It is thought that when a nestling swift leaves the nest on its maiden flight, it may not touch ground again until it itself breeds at two or three years of age. It would be as if a human baby on taking its first steps continued walking non-stop until it was something like seven years of age.

Echolocation: Most of us equate echolocation as a purely mammalian technological feat. Ignoring our own rudimentary abilities with radar and sonar we all know that bats emit high frequency sound pulses and use the returned echoes to build a picture of their surroundings. Likewise dolphins have a specialised fatty head structure called a melon which allows them to interpret reflected sound and to know what is in front of them; a distinct advantage when swimming through murky sediment rich waters.

A handful of bird species have also mastered the art of blind sight, including a tribe of swift species known as the swiftlets. Species belonging to this group build the nests in dark caves, and finding the way back to their nest in the darkness is a considerable challenge. The swiftlets have solved this problem by developing a form of echolocation. The swiftlets emit short clicks, distinctly audible to man. It is thought that by gauging the time it takes these to return the swiftlets can work out where they are in the cave and avoid bumping into the walls.

Adult and nest.

Sticky superglue: Spiderman would be proud of the swifts. As they build the nests in usually elevated locations there is a great danger that loosely built nests might fall considerable distances. Nests need to be well anchored to avoid this distressing occurrence. Swifts have solved this problem through producing one of the strongest glues known in the biological world. Nests are literally welded together with particularly strong glue made from their saliva. Once made nests are resistant to the elements for years.

Long migration: Despite the fact it spends most of the year in Africa the swift is often known as the European swift. It would actually be more correct to say it is an African species which simply takes an extended holiday in northern Europe for child rearing purposes. Swifts migrate from southern Africa to Europe each April, covering a distance of over 5000 kilometres, only to spend an average of 100 days here before starting the return journey at the beginning of August.

The swift can rightly be said to be a long haul flyer. The fact that swifts migrate was long controversial, with it being widely believed that swifts hibernate through the cold winter months. Like much folk-law this idea might contain a grain of truth, as swifts can enter a form of torpor in cold conditions. The first swifts return in the spring on around the first of May.

These first arrivals are followed by the main body of swifts about a week later. However, when exactly swifts return is dependent on the weather. When conditions are good swifts arrive several days earlier, likewise bad weather delays their arrival. The swifts begin their return to Africa at the beginning of August when the young have successfully left the nest.

High octane fuel: Flight of course is an energetically expensive activity, even for an efficient flyer like the swift. Like a fighter jet the swift requires a form of high octane fuel. But are there enough insects for the swift at the high altitudes at which it flies? When we think of high altitudes we tend to think of the empty cold skies seen though plane windows. Actually swifts find plenty of food. Swifts feed on protein and energy rich insects such as spiders which they grab from the air as they swoop past with their mouths wide open.

Nesting behind waterfalls: If you thought that only humans were so stupid enough to engage in extreme sports, then think again. A number of swift species, those belonging to the Cypseloidinae subfamily living in South-east Asian, and known appropriately as the waterfall swifts, like to nest behind waterfalls. In fact many of these species are dependent on such places. Of course the great advantage of nesting in such locations is that they are predator free. However, rather obviously, there is one great disadvantage. To reach the nest the adults have to fly through a streaming torrent of water. This does not seem to be a problem however.

Chapter 3
First Contact

The first time I met the swifts was in April 2007. The previous month had been warm but there had just been a short cool spell of weather. It was not really bad weather, simply chilly. Then over a few days the weather became warmer and more summer like. The sun shone, there was no wind, and it was very pleasant.

A few days later I noticed the arrival of the first swifts of the summer. I cycled to work every day along a cycle path which for some considerable way followed a river. I knew from previous years that this stretch was a favourite hunting ground of swifts. They would fly above the river catching the insects which emerged from the water or the banks. I was cycling along relatively early in the morning thinking about all the things I had to do that day. I scanned the sky as I normally did to check for swifts, but without really expecting to see anything. It was still only the middle of April, and the first swifts should not have begun to arrive before the beginning of May.

I was surprised to suddenly see twenty or more swifts swooping and curling in the air above me obviously trying to catch food. I simply stopped peddling and sat looking up at them for about half an hour before continuing with my journey.

The way into the bridge.

From then on I visited the bridge each evening to see if any swifts would start to stay in it overnight. Not wanting to disturb the birds in the chambers. I watched from outside on the reservoir banks. Here I could see underneath the bridge fairly well. Although the swifts had by now been in the area for several days, there was no evidence that they had been to the bridge. They probably spent the night on the wing instead.

Then after five days of fruitlessly waiting around the bridge at dusk, a few minutes before darkness fell, and without any hesitation, a single swift flew from the south and dived straight into a hole. I had expected the first arrival to maybe fly around the bridge several times, or to fly around the nest holes searching for a suitable one. But it had simply flown directly in. Maybe it was an experienced bird that had used the bridge as a nesting site in previous years. Within a few days I noticed that several birds were now coming to the bridge at night to rest. There were several piles of fresh droppings around some holes.

The bridge apppeared empty during the day. Then we had a rainy spell.. For three days it did nothing but rain. I saw no swifts flying around and figured that they had probably flown to Italy to avoid it. The bridge would probably be empty, or so I thought. So I entered without much fear of meeting a swift. It was only when

I went into the second chamber and inadvertently shone my head light at a nest that I saw two pairs of beady eyes staring back at me from only half a metre away. I scanned the chamber and saw a dozen or so other eyes all looking at me curiously. The birds were sat either in the nests or by the side of the holes.

Some birds were in pairs, but there were also many who were alone. When I shone my head light on the birds some started ruffling their feathers. A few crept over to the holes as though making ready to head outside to safety. One flew out. I decided to beat a strategic retreat to avoid disturbing them.

Instead of flying to sunny weather they instead had used the bridge for shelter during the bad weather. The swifts were obviously more sensitive than the Swallows and House Martins, as these two species continued to hunt and fly above the reservoir surface despite the heavily falling rain. The swifts however had decided to rest in the warmth and dryness of the bridge.

Beginning research

The first year I studied the swifts the weather during the spring was most unusual. Spring had come early. In April the weather was fantastic. The temperatures were more like one would expect in midsummer in July than at the beginning of spring. As the swifts had arrived two weeks earlier than I had been expecting, I thought that the whole swift season would occur two weeks earlier than usual.

The shock came in the first week of May. It turned suddenly cold and it began raining pretty much non-stop for the following two weeks. I began fieldwork in the middle of this cold snap. Visiting the bridge it became apparent a couple of birds had jumped the gun regarding egg laying. They had obviously started laying in the warm spell only to be surprised by the following poor weather.

One nest contained three eggs, but on subsequent days no more were added and incubation did not begin. They had obviously been abandoned by the pair. At another nest a single egg had been laid but this had been ejected from the nest. The other swifts were obviously delaying pairing and egg laying until conditions improved.

Swift courtship

At this time it was not unusual to see groups of 30 to 40 swifts flying around the bridge. They would normally fly in loops and circles over and around the bridge, occasionally flying beneath the bridge arches in acrobatic show-off play. Sometimes I would see pairs of birds playfully circling one of the struts time and time again with one of the pair seeming to chase the other.

Nests were often close to the holes.

Birds entered holes, only to re-emerge after a few seconds. Often several adults would visit a hole in turn. To me it looked as if the swifts were 'house hunting', checking out potential nest sites for their suitability and availability. On one occasion a swift tried to enter a hole in which there was already a swift inside. The swift inside began calling out loudly and repetitively until the intruder had retreated. The calling continued for a number of minutes even after the intruding swift had left. This was more evidence of the swifts possessive nature in regard to their nesting sites.

In David Lack's book, 'Swifts in a Tower', which describes his life with the Oxford swifts, David Lack described a form of behaviour sometimes seen in swifts which he named 'banging'. This was where several 'foreign' swifts from another colony would fly low over a nesting colony in close formation screaming loudly. David Lack speculated that these foreigners were probably searching out vacant nest sites. However, this does not really make sense. Swifts nest in the same roost year after year and so do not need to find new nest sites each year unless they are juvenile birds. Although I had seen several swifts flying in such a close formation close to the drainage holes, they appeared not to be foreign swifts but rather swifts that were already members of the colony.

It was often the case that the swifts would fly around the bridge for half an hour or so, then suddenly seem to disappear completely, probably going off to feed. They would only return two or three hours later for another session of play. This pattern would be repeated throughout the day, with their being short periods of frantic swift activity at the bridge, followed by longer periods when the swifts flew away and disappeared from sight.

I knew that when there was splendid weather then at the bridge there would be much swift activity. The swifts did not have to struggle to find food and therefore maybe had some time for acrobatic play.

Getting to know each other
Naturally as the spring progressed both I and the swifts began to spend more time at the bridge and therefore came into contact with each other more

often. When I was inside the bridge and there was a swift nearby they would often 'freeze' when they sensed I was nearby. They would simply remain motionless. I could even reach out and touch them.

On one occasion when I was walking past an unoccupied hole where there was no nest, an adult swift flew up through the hole. However it had obviously become aware of my presence before completely getting into the bridge. It froze half in and half out. Its feet were resting on the edge of the hole, its head was inside, and its body somewhat outside the hole. It hung like this completely motionless for several minutes, looking upwards without moving, before eventually summing up the courage to disappear back outside again.

Other birds were more distrustful of me and more flighty. As soon as they became aware that there was a foreign presence in the bridge they would dash to a hole and fly out. If I spooked a swift when it was not expecting it they would sometimes even take off within the chamber and fly around in ever decreasing circles until they flew into a wall and slid to the chamber floor.

Then they would often crawl to the nearest hole and fly quickly out.

I tried my best to avoid such occurrences as it was obvious it caused undue distress to the swifts. As the summer progressed birds became more tolerant of me and appeared not as scared. They even began to scuttle over towards me on occasion. At first this behaviour puzzled me because you would really expect them to scuttle away. It turned out that what attracted them to me was my head torch. If a swift was scuffling over towards me and I turned my light off it would stop and instead go towards a hole.

Of course this behaviour makes perfect sense to the swifts. Within the bridge normally any sign of light comes from a hole and is a way out. The swifts thought my lamp was just another kind of hole. As my head torch was several times brighter than the light from outside it is probably not surprising that the swifts were more attracted to it than the light from outside. You could not really blame them for being confused. On these occasions they would often circle around me and even crash into me obviously attracted by the light shining from my head.

Some individuals proved quite proficient fliers inside and could fly several circuits within the chambers without hitting the walls. This made me wonder whether they had any bat like ability to see where they were flying in the darkness. When a swift flew at you in the dark the most noticeable thing about it was its bright white chin patch, which shone out well in the darkness. I tested with a bat detector in the bridge but found no evidence of low frequency calls.

I was still unsure of how much handling the swifts would tolerate. At the beginning of spring I had caught two adults from one nest and taken their measurements. Afterwards I had left them on the nest and they seemed to settle down well. When I went back a short time later they were still where I had left them, as they were in subsequent days. I therefore concluded that the swifts were relatively resistant to my handling them.

So a few days later I caught another swift at another nest and again took measurements from it. Again after I had replaced it back on the nest it seemed to settle down and did not fly off. However when I checked the next day, the eggs were uncovered and there were no swifts near-by. The parents had deserted the nest. After this I decided not to catch any more adults until the young had hatched. They were simply too sensitive.

On several occasions I came across adult swifts hanging onto or climbing up the walls of the chambers. Sometimes they would fly about and then come to rest on the walls and remain suspended there. The strong legs and claws allowed them to remain apparently hanging on completely flat vertical surfaces without effort. One bird remained hanging around, so to say, for over half an hour.

Swifts have often been depicted in pictures hanging onto walls. David Lack pointed out that many of these pictures are false in that they show suspended swifts twisting their heads so to look to one side. He said this was not really possible for the swift because its neck is so short and its head is so closely associated with its body. However, I can confirm that swifts do hold their heads to the side. These hanging swifts manage to look to the side of them as often depicted, despite being unable to twist their necks.

On one occasion while walking through the chambers I came across two birds struggling with each other. They were locked together by the feet. At first they appeared motionless, but every few seconds they seemed to twitch and struggle with each other. They were both laid out on their sides with their wings outstretched. They both appeared quite exhausted and both had their eyes closed. They were oblivious to me and the light which I shined on them.

I noticed that one of the birds seemed to be fairing worse than the other as its belly and legs were covered with blood. I assumed its opponent had caused these wounds with its sharp claws.

How the fight could have started I didn't know as both birds were relatively far away from a nest or entry hole. I disengaged both birds and on becoming unlocked they seemed to become suddenly aware of me and both dashed through a hole and flew away.

A climbing swift.

Why they should have been fighting I don't know as chicks were hatching out at this time and it was too late to start a brood now. I guessed that they were two males, one of which had intruded onto another's nesting site. The mystery of the fighting birds was soon solved. I marked one with a large white spot on the head before it flew off. I was later surprised to find this individual sitting on a nest in a neighbouring chamber. I can only assume it had entered the chamber by mistake and thus become engaged in a fight with the occupant.

Opportunities for such observations often occurred when I was in the bridge doing something and birds flew in around me. I would immediately turn my head light off to avoid scaring them, and sit down to watch what they would do. Most of the time they would do what one would expect a resting swift to do, namely preening and ruffling their feathers.

Occasionally birds would 'cheep' softly, seemingly to themselves. When other swifts flew under the bridge they would sometimes become extremely animated. Some birds would stand right on the edge of the hole peering out as far as they could and call back with a harsh and loud 'chi-chi'. These seemed to be replying and calling out to the swifts outside. When these disappeared they would settle down again, but later when more swifts flew by, calling would again be resumed. Were these males calling for mates? Or were they signalling possession of the nest holes to potential rivals?

I also noticed that swifts were calling to each other within the bridge. A swift at one end of a chamber might start calling out, and then another swift on another nest at the far end of the chamber might reply back. If you stood outside the bridge sometimes the air around would ring with their 'chi-chi' calls although none of the swifts were airborne but all in the bridge.

Mostly the birds would simply sit peacefully on the nests. I noticed that usually birds that were in pairs would sit close to each other quietly. The birds were quite possessive of their nest sites and the areas surrounding them. Some drainage holes were situated widely apart, but others were only seperated by 30 cm or so. I watched two swifts at different but closely situated holes facing each other and occasionally emitting their distinctive 'chi-chiing' to each other. Suddenly one of the swifts shuffled along the floor of the chamber in the direction of its apparent rival in a kind of challenge. When the swift saw its challenging rival it left immediately and flew off. The remaining swift then proceeded to shuffle in a two metre wide circle around the holes, as though checking the other swift had definitely left and not just slunk into a corner or out of sight. This swift in its aggressiveness seemed impervious to the light I shined on it.

One of the favourite occupations of the swifts inside the bridge was peering outside the entry holes. You could see them cocking their heads to one side to get a better look at something of interest down below, be it some ducks, a boat, or simply the glimmering water. At the start of my fieldwork I usually saw birds alone in the bridge, but as the weeks wore on they were more commonly in pairs.

Partners would often preen each other's head and throat, and this probably helped to strengthen the pair bond between them. Sometimes the swifts would yawn in a very comical manner, and the wide opening gape of their mouth became apparent. Although when they moved along the floor they had to 'belly-flop', they could nevertheless move at considerable speed. One bird travelled a circle of a few metres in diameter around its nest extremely quickly.

Despite being such elegant flyers the swifts often seemed reluctant to make the final leap through the drainage holes and into the sky. When I had scared the swifts by accident they seemed much less hesitant about leaving. If they sensed my presence they would simply drop out.

Swift re-entry.

You would think because of the apparent difficulty swifts have at moving over the ground that they would place their nests as close as possible to the entry holes. However this does not appear to be the case.

Few swifts nest directly next to entry holes in the bridge. Most place nests 20 to 30 cm away from the entry holes, and in some cases even farther away. Sometimes the nests are placed on elevated ridges which the swifts then have to ascend each time they want to reach the nest. Maybe the mode of locomotion on the ground appears more difficult than it really is. Or the advantages of having a more sheltered nest outweigh the difficulties in having to reach it.

I was acutely aware that other projects studying the swift had been beset by the problem of the swifts deserting the nest when they had been disturbed in some way or other. I did not want this to happen to my study. I wanted to use an overly softly-softly approach inside the bridge, even if it meant missing out on some of the data that I wanted. I did not want to lose a unique and special study site because of a few acts of clumsiness and heavy handedness.

The swifts continued their show-off acrobatic flying for about two weeks. At the end of this period they began nest building.

Chapter 4
Nest Building

During the nest building phase I often stumbled across birds either on nests or near to holes obviously engaged in nest maintanance. Many seemed to be coming and going all day.

Swift on the nest.

I stopped and observed one newly arrived bird rearranging the nest to its own satisfaction. It had flown in with a beak full of moss and then had spent 10 to 15 minutes shuffling about on its nest and moving the pieces of moss about until it was happy. It then flew off, presumably to collect more nest material. Many of the birds seemed to be likewise occupied.

When working on their nests birds showed a distinctive pattern of activity. They would arrive at the bridge with a beak full of nest material. They would then sit within the nest or at one side and begin weaving it into the walls. When building new nests the swifts would start with the rim. They would first make a wreath like circle, only later adding the bottom. Presumably the rim was the most important part of the nest as it stopped eggs from rolling away or getting lost.

A few pairs were building new nests, but most were adding new material to existing ones. In other words, simply 'renovating' them. This new material was mostly composed of grass, straw or leaves, but the influence of man was evident with string, plastic, and even silver foil also being used. The swifts would weave this together and stick it together with their super-glue like saliva. The old nests were in most cases rather hard and solid structures. Many were stuck firmly to the ground.

At some of the holes where there were no nests some swifts started building new ones. I assumed these were juvenile swifts, They did not breed or lay eggs but worked on preparing for the next year's breeding attempt.

A clutch of white eggs. Note the parasitic Louse-fly.

That this was occurring was obvious because green material kept being added to some nests long after the other pairs had young. At one hole a swift had begun three 'circles' of a nest, but had not got very far with any of them. I wondered if this was a reason why it paid juveniles to wait a year or two before breeding. It allowed them to practise nest building and egg laying or to find a nest site and build a nest.

At two of the pseudo-nests eggs were laid, but no attempt was made to incubate them. These eggs were laid much later than the other clutches. In one case they were laid well after many other clutches had hatched out. Adult swifts first breed in their third or fourth year. So this nest building could be considered a form of work experience; the birds practise and hone their nest building skills before beginning to breed seriously the following year. A kind of swift 'gap year'.

It must be remembered that a swift summer is more or less a race against time. The eggs are incubated for a relatively long period of time compared to other species. The young have to be reared for over a month before becoming fully independent. The swifts return to Africa at the beginning of August. They therefore have to fit their breeding into only 100 days. Time is of the essence. Adult swifts do not have time to spend searching and choosing a partner, or in practising to build nests. They need to hurry up and get on with the important business of incubating. Having a gap year could be an important adaptation in helping save time in future years. I would not be surprised if some of the pseudo nests get developed into proper nests and used in future years to save precious time.

This lack of time probably also explains why swifts form pair bonds for life and are nest site loyal. They simply do not have time for mucking about finding a new partner or new nest site each year. It makes sense to make a nest and stick with it.

Swifts were extremely possessive over their nest sites and as a result only one pair of swifts used each hole. In theory it would have been possible for two pairs to use a single hole and to build their nests some distance away from each other.

The number of nests in different chambers varied considerably. In one chamber there were no nests at all, while in another there were seven. Within chambers nests were often situated at adjacent holes, or opposite one another, thus being close together although distant at the same time.

The swifts apparently liked to nest in close proximity to one another. But not too close. This sort of sociality makes good sense biologically. By nesting with others you reduce the risk of predation to both yourself and your offspring. Any potential predator can only attack one nest at once. For a solitary breeder the chance it will be you is 100 percent. If you breed in a group of 10 the chances you will be the one attacked are reduced to one in ten, all things being equal.

Of course nesting in groups makes yourself and your group more conspicuous simply because there are more of you, but the benefits must override this disadvantage. And if there are already others nestling at a location it must be a fairly safe place to nest otherwise they would not be there. Unless of course disruptive biologists come to visit.

Egg laying

In David Lack's classic study of the swift in the 1950's he had noticed that swifts began laying their clutches of eggs exactly five full days after the onset of a spell of good weather. Swifts lay typically between one and four eggs. There is usually a gap of two days between the laying of each egg. The female will continue to leave the nest to feed, only beginning to incubate the clutch with the laying of the third egg.

Therefore for the most part swift researchers have been able to record and measure the first two eggs, but not those laid thereafter. Swifts are too sensitive to be disturbed once incubation has begun. The timing of laying of a fourth egg, when this occurs, is pure guesswork, but probably occurs two or at most three days after the third egg appears.

It was interesting to note the number of pairs on nests. Most sat side-by-side facing each other. Only rarely would a bird be alone without a partner. This was in stark contrast to only a few days earlier when many of the birds had been singletons. This reminded me of prospective fathers who stayed with their pregnant wives during labour in order to see the birth. It seemed quite touching. But then my more pragmatic biologist side kicked in. I knew there was probably a very good reason for this closeness between the pairs at this time. The females would soon be laying eggs and the males who mated with the females would end up as fathers. It would pay the males to stay with their partners as much as possible to ensure that they and only they mated with the females.

What I was seeing was probably a form of mate guarding. The males were not staying with the females out of love. In the battle to ensure genes get passed it pays to keep your partner under close supervision.

However, if you watched the pairs for any length of time they did seem to show affection to each other. They would often preen each other's heads and throats. Or sometimes they would mutually look out of the holes together. When other swifts flew underneath the bridge one of the pair would normally call out, maybe warning potential rivals to stay away. As well as noticing that the birds seemed to have formed pairs within the bridge, I also noticed that there was much less group flying going on outside the bridge. The frantic play flying of the previous days was replaced by much more practical a-to-b flying. Now when I saw swifts flying it was not in large groups, but rather in twos or threes. Again a form of mate guarding?

One of the reasons I was so surprised to see the swifts in the bridge was because I assumed that the making and laying of eggs must require a considerable amount of investment on the part of the female. If each egg weighed on average 3.5 grams, and she laid an average of 3 eggs, than in total she would need to produce roughly 10.5 grams of protein over only three days.

To illustrate how much of a feat this is consider a human equivalent. A mother would have to lactate 15 to 20 litres of milk for her baby in only three days to match this amount of expenditure. Producing the eggs is an energy and resource sapping business

I expected the females to spend as much time as possible fuelling up before they began laying the eggs. This was obviously not the case. Maybe instead the resources were built up over the whole five days of good weather and the female birds could afford to lose a day's hunting before they laid the egg. Whether this lack of feeding is forced upon the female by the mate guarding male or whether it suits the female to have a rest is unclear.

The great egg lay occurred on a single day! One day I was late getting to the bridge. It was mid-afternoon before I was there, and there was unusually not a swift in sight overhead. I tip-toed with great trepidation into the first few chambers. I expected, like yesterday to find swifts around many of the holes and on the nests. However it soon became apparent that apart from a solitary bird, the bridge was devoid of the presence of birds.

I was over the moon to find a single small all-white egg in nest after nest. There was one in almost every second nest. This was really something! I crouched down and undid my callipers and table scales to begin measuring and weighing the first egg. My hands were shaking with excitement. With the discovery of each new egg I became more and more happy. The birds were a day late according to David Lack's five day theory, but they were quite close. The amount of synchrony throughout the colony was amazing.

There were some differences between the size and weight of the eggs. The lightest egg was three grams, and this had been unceremoniously dumped outside of the nest, probably because it did not meet the strict standards the parents expected it to have. The heaviest egg weighed in at 3.8 grams. Most averaged about 3.4 to 3.6 grams. In size the eggs were 22 to 25 mm long, and had a circumference at the widest point of between 15.9 to 17.2 mm. Although a total of 12 clutches had been started on this first day, I hoped for more the next day.

Interestingly enough, although in this first year the swifts had arrived some two weeks early because of the good early weather, the egg laying itself did not begin any earlier than as normally reported by David Lack and Weitnauer. This suggests that although the arrival of the swifts is influenced by weather conditions, the timing of egg laying is instead influenced by day length. Although weather conditions might cause a delay or early start to egg laying, this is typically of only a few days, and egg laying takes place at roughly the same date each year.

Egg laying had obviously continued the next day. I got to the bridge around midday. There were a few swifts flying about outside but it seemed fairly quiet. I entered a chamber, and soon noticed a pair of birds sat at their nest a few feet in front of me. I turned off my light and quietly crept around them as best I could and went into the next chamber. The same story here; again there was a pair of birds on a nest. The picture was the same throughout the entire bridge. I was able to check a few nests at one end but then had to proceed in darkness. I checked several more chambers but they all contained quietly sitting birds and this made my work more or less impossible.

In the several nests I was able to examine I saw that several new clutches had been started, as a single egg was present, whereas yesterday there had been none. However, at none of the nests where yesterday there had been a first egg was there a second.

An adult brooding newly hatched chicks.

According to David Lack the second egg of the clutch is laid two days after the first is laid. So in theory those swifts that had started clutches yesterday should lay their second egg tomorrow. The laying of the third eggs a few days later signalled the end of my current phase of field research.

Although I had measured and weighed a number of the second eggs in each clutch, there was practically no way for me to measure more, or to measure the third and even maybe the fourth eggs which would be laid without disturbing birds. It is known that swifts will leave the nest uncovered for a short period each day, roughly of about half an hour to an hour, but to visit the bridge in the hope that all the birds would take this 'break' simultaneously at the same time allowing me to study the clutches was an unrealistic hope. I would not be able to work my way around all the nests in such a short time period anyway.

I just had to take a frustrating enforced holiday from the bridge and the swifts and come back again just before hatching. In this first year I was able to measure 33 'first' eggs, 8 'second' eggs, but only 3 'third' ones.

Hatching

I stayed away from the bridge for two weeks until I felt I could again enter the bridge without disturbing the birds unduly. Incubation lasts exactly 21 days in swifts. Typically the young begin to hatch out about the middle of June, although this depended on weather conditions and could be slightly earlier or later.

Newly hatched chicks. Note that this nest is built in a layer of green moss.

The first time I saw chicks was when I checked a nest near the entrance. I could see that under the parent the chicks formed a wriggling ball of flesh. As I approached nearer, the adult stepped off the nest, but did not retreat down the entry hole, instead remaining motionless a few centimetres from the nest. It did not seem to mind when I stretched a hand out and began taking the chicks one by one from the nest.

The chicks were revolting. They are naked at birth and had pale pink skin all over their bodies. The head and neck are small and delicate looking, but the abdomen is large, a little like a ball, and in proportion to its size the legs are strong and long. The eyes are closed, but visible through a closed layer of skin through which they are black in colour.

It was relatively easy to work out on which nests chicks had hatched, despite one of the parents always being present on the nest and thus concealing the brood. Although you often could not actually see the chicks under the parent bird you could often hear them. The chicks gave out a short high peeping sound that was noticeable some distance away. They would continue to peep even in your hand. It normally pays chicks to be quiet in the nest to avoid attracting unwanted attention from predators, but maybe for the swift with its inaccessible roosts the pressure to be quiet is not as strong. However after a day or so the chicks did learn to keep quiet, only peeping out when particularly hungry or when a parent arrived with food.

Another dead giveaway that chicks had hatched was when there were pieces of eggshell around the nest. The chicks had either pushed these out as they hatched or the parent had placed them there. Later the parents would remove the eggshells.

When sat on chicks as opposed to eggs the parent would appear to sit rather uncomfortably on the nest and fidget and move about. Having wriggling chicks underneath them probably did not make as comfortable a cushion as the eggs did. When it was particularly warm the adults would not sit on the young, instead rather standing over them. In such cases it was probably warm enough that the young did not need any additional warmth, but the parents still had an instinctive urge to remain on the nest.

Around the hatching time I was consumed with worry. I wanted to know on what days chicks hatched so I knew how quickly they put on weight for my research. This meant I had to walk through the chambers at least once a day. What if the parents deserted? I had stayed away all through incubation. What if the adult swifts became too startled by my presence and left the nests never to return? It would be a shame to disturb the birds and it could mean the end of my research. This meant I was extremely cautious at this stage. I moved with the utmost care. I swiftly retreated if birds appeared upset or startled by my presence. I often 'missed' nests to avoid undue disturbance.

Chapter 5
Young Nestlings

As David Lack pointed out a newly hatched out swift is superbly adapted to life in the nest. At this point in its life it has only to do two things; eat as much as possible and stay in the nest. Its large mouth gape is already in evidence at this stage, and the large abdomen acts as a dual food processing unit and fat store.

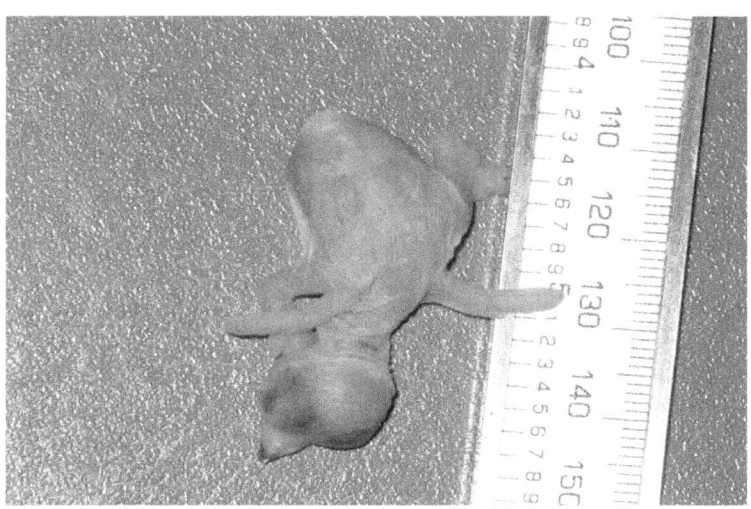

A newly hatched chick.

The strong legs and claws allow it to keep a firm grip and stay in the nest. A nestling outside of the nest will be ignored by its parents and will quickly die unless it can get back inside. The massively engorged abdomen is similar to those permanently upright baby toys which are often in the form of a clown. This toy has a rounded and heavy bottom. You can tip the clown over but he always rights himself because of the low centre of gravity and the round bottom.

Swift nestlings seem to have evolved a similar method to help them stay in the nest. The weight is centred fairly low down plus they have really rounded bellies. If you hold a swift nestling in your hand it will remain in the same place even if you try to tip it away. This is probably a useful adaptation giving the nestling the necessary ballast so that it remains anchored in one spot.

When first hatched out the chicks are naked and their skin is a dark pink colour. After five or six days dark spots can be seen under the skin of the back. This is where the first feathers would sprout from. From about the ninth or tenth day of their lives, coinciding nicely with the end of brooding by the parents, small downy tufts appear on the back. A few days later small tubes sprout on the wings and head. From this age the nestlings presumably begin to produce warmth themselves.

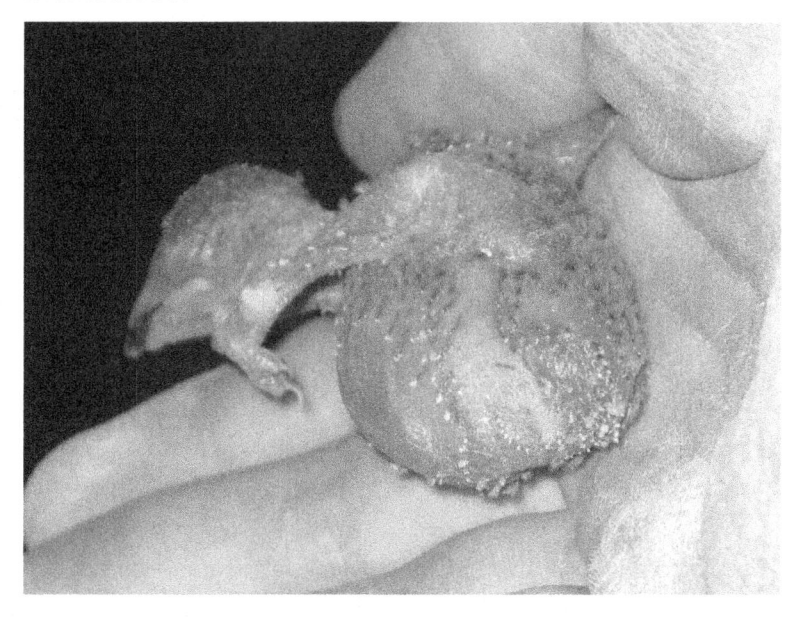

Nestlings have big bellies.

These small tubes are the beginnings of the real feathers and after a few more days opened up completely. At this time the nestlings were if anything grey in colour, that being the colour of the rolled up tubules. When the feathers finally emerged the nestlings became black and they therefore took on a much darker colour. At first the black colours were restricted to the head and wings and tail, but this soon spread over the entire body.

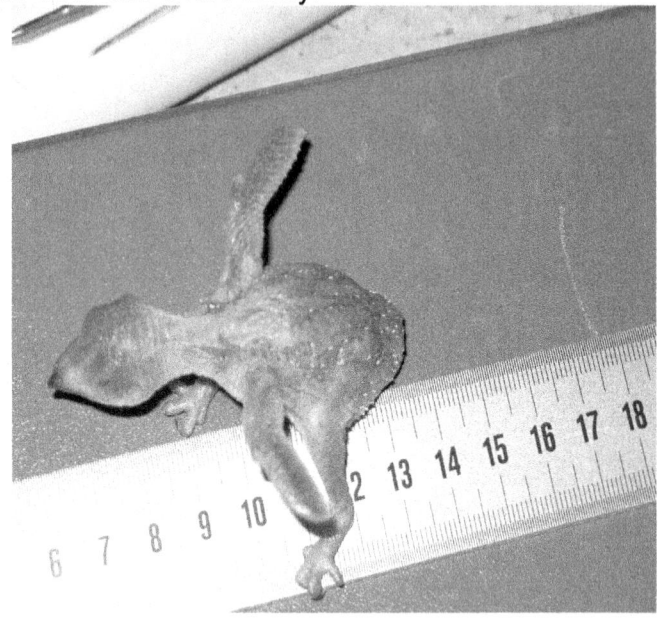

Nestlings take on a grey colour when the first feathers sprout.

Nestlings seemed to instinctively snap out with their mouths, especially if I touched them around the head. This reaction became more pronounced as they got older. Nestlings of a week or so old would eagerly snap at my pencil or callipers.

The nestlings have long thin necks, and these regressed with age. The adults have hardly any neck. These necks probably enable a nestling to put its mouth as close to the direction food arrives from as possible. They also mean the head can be positioned and twisted towards food, a bit like the hose on a vacuum cleaner. Moving their entire bodies around is problematic due to their heavy abdomens. When particularly hungry the nestlings would throw their heads back and open their mouths and peep for food.

Although the nestlings cheeped and peeped in the nest at regular intervals, they burst into a chorus of begging when an adult returned to the nest with food. This was especially the case when they were older or there was more than one nestling in the nest.

A major problem I faced was in marking the chicks. The skin must be particularly oily and this must act in a self-cleansing way. Permanent marker disappeared within a day. Tippex was gone within two or three days. In the end I simply had to try to tally the weights and measurements from day to day to work out which chick was which.

The nestlings soon began putting on weight, and indeed very quickly. I weighed and measured them each day. Sometimes this was easy to do as the brooding parent would disappear as soon as they saw me coming. This allowed me to do my work without having to worry about the adult flying around or panicking. On other occasions the adult would remain on the nest and I would have to fish for the young underneath it. It would just sit there while I gently moved it to one side and pulled its babies away. On other occasions the parent would leave the nest and remain close by as I worked.

The worse scenario was if the adult panicked and began fluttering around or flying within the chamber. Then I would turn the lights off and hope it disappeared or settled down. I always tried to move thoughtfully and slowly around the birds making as little noise as possible. I was quite lucky in the respect that I had always kept birds all my life. If you are experienced at handling birds you seem to instinctively develop slow thoughtful movements which don't startle or surprise them. I have often seen beginners grab out wildly and then wonder why a bird is startled, while I could easily touch a bird while it sat perfectly calm.

The easiest way to weigh the chicks at this early stage was to tip them upside down and place them on their backs on the scales. Although they would struggle madly to right themselves and wave their legs wildly in the air, this ensured they stayed relatively stable on the scale top. If weighed the right way around they would scramble to get free and often fall off. Later I started using a small plastic tray into which I placed them in so that they stayed in one place.

Weighing a young chick.

I noticed that often after being placed on their backs the chicks would do a poop, normally in my hand. I wondered whether this was some kind of natural reaction and whether the parent birds either tipped the small chicks over or more likely stroked their backs to induce them to defecate. This would allow the parents to easily keep the nest clean and allow them to clean up after the chicks in a controlled manner.

The anal opening of chicks is rather strangely located. Instead of being where you would expect it to be, somewhere underneath and to the posterior of the chick, it is in fact on top of their rear. This allows the parent to easily remove the droppings produced without having to fish around under the nestling. The anal opening moves to the more usual posterior position as the nestling ages.

Typically nestlings grow and put on weight rapidly. From a starting weight of about three grams at hatching they reach a weight of 50 grams roughly 30 days later. This means they put an average of 1.56 grams of weight on each day they are in the nest.

Parents deliver food bundles weighing around 1.5 grams on average eight times a day. Thus they provide roughly 12 grams of food each day. This means that 14% of the weight of food the nestlings receive gets turned into body weight. This level of efficiency is impressive when you consider such small animals probably need to expend considerable amounts of energy on keeping warm because of the large surface to volume ratios of their bodies, and the lack of insulating feathers or fur.

Nestlings have a cheeky appearance.

The graph on the pages following shows the weight changes in four different chicks from four different nests from my first year of fieldwork. It is purely to illustrate weight gain. Two of the nestlings survived and successfully fledged and left the nest. The other two nestlings died before reaching maturity.

As you can see from the graph chicks sometimes have days where they do not gain any weight and even periods when they lose weight. These periods correspond to days when the weather is bad and there are not many insects flying around. These could be days when it is simply too cold for insects to be around. But it could be simply that parents can't hunt because of heavy rain or wind.

Although initially all four nestlings gained in weight, the weights of the chicks which died quickly levelled off, while that of the successful nestlings continued to increase at a fairly rapid pace. The chicks which died lost a significant proportion of their weight before dying.

For example the third chick's weight dropped from nearly 30 grams on its 15th day alive to around 20 grams at its time of death. In effect it lost 50% of its body weight before dying. The successful nestlings on the other hand, continued to put weight on continually for most of the time they were in the nest.

Continued weight gain the key to success

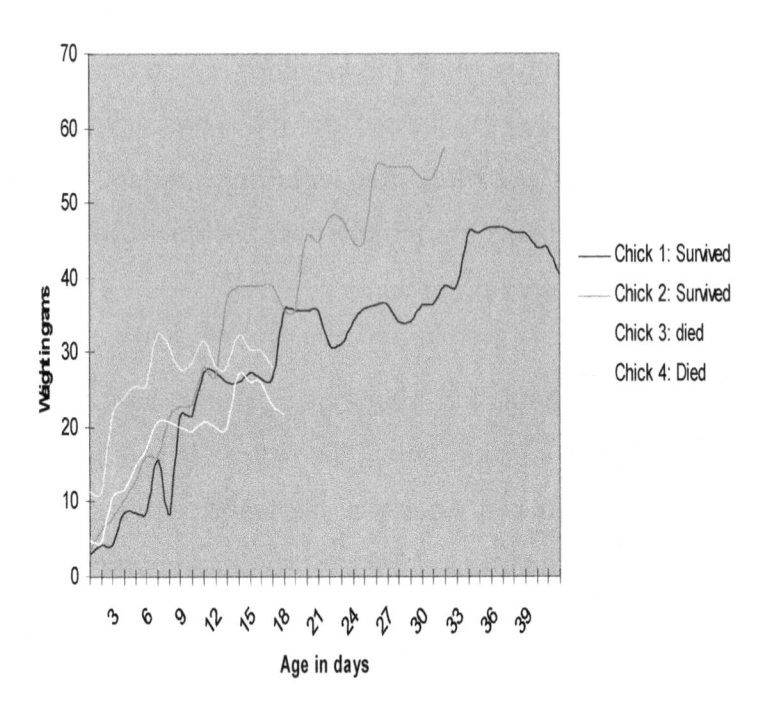

- Chick 1: Survived
- Chick 2: Survived
- Chick 3: died
- Chick 4: Died

That the young swifts can take such large drops in weight before finally succumbing is also quite impressive. I weigh 63 kilograms. If my weight dropped as low as 50 kilograms I would be seriously ill and at great risk of dying. This would be a weight loss of only about eight or nine percent of my original body weight.

Wing growth is continual and steady

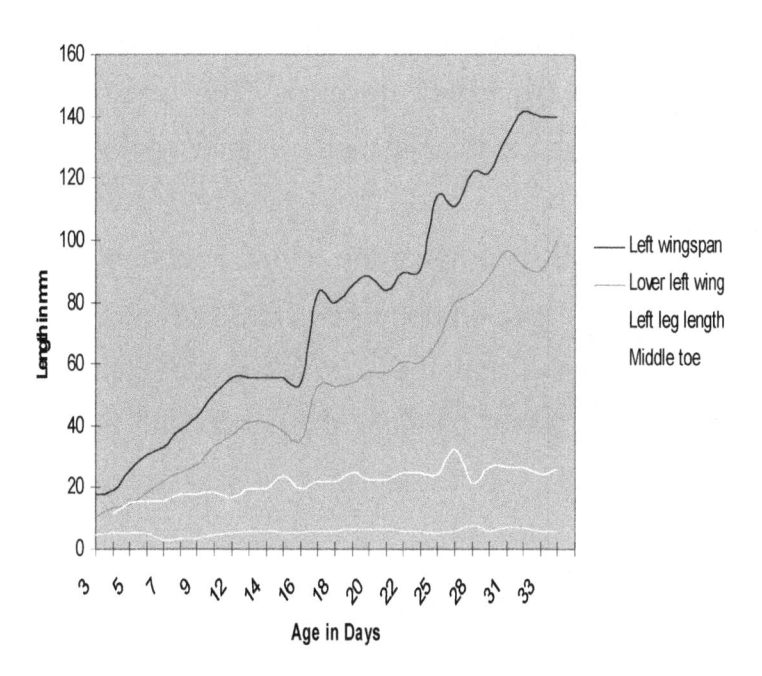

[Graph showing Length in mm (0-160) vs Age in Days (3-33), with four lines: Left wingspan, Lower left wing, Left leg length, Middle toe]

Nestlings on the other hand can lose 50% of their body weight before dying. Nestlings can easily lose 20 or 30% of their body weight and still recover without any problem. However the comparison with me is not a fair one. Nesltings are still growing and developing while I am fully grown. Nestlings in the nest have evolved to become little 'fatties' as this provides the resources needed to overcome poor weather and starvation.

However, weight is not the only factor to consider when studying how nestlings grow. The second graph shows how wing length develops. The length of the wings grows more steadily than weight is gained.

The wings are perhaps the most important bodily feature of birds for an aerial specialist like the swift, so it makes sense that these receive priority and are grown steadily and carefully throughout the period of development.

Another measurement I took was the left middle toe length which is pretty much a standard measurement taken by ornithologists when studying birds. However in the swift this does not change much during development. Typically these stop growing from around 10 days of age, and reach a length of only seven millimeters at most.

The same goes for the left leg length, which reaches a length of around 20 mm also by about day 10. The claws and feet are disproportionately large in newly hatched and young swift nestlings, but this makes sense, as this is the time when they are of most importance to the nestling in keeping it anchored into the nest. Later they are not as useful and while other parts of the body grow they remain the same in size, and so become proportionately smaller in size.

Previous researchers have noticed that nestlings seem particularly resistant to cold. I often observed parents leaving nestlings unattended for several hours at a time. The nestlings quickly cool down and become cold to the touch, but this does not appear detrimental in the least. Whereas nestlings of other bird species die within an hour of being left by a warming adult, swift chicks seem immune to such cooling and although they may appear lifeless they quickly pick up when warmed by a returning adult. In cases where adults deserted the chicks completely, never to return, the chicks would remain alive for a staggering three to five days without receiving any external warmth at all.

My nestlings also seemed fairly robust and resistant to cold despite their small size. Coldness did not seem to affect their level of activity or begging behaviour, at least at first. In your hand the chicks would rarely sit still, they would be constantly moving and wriggling and begging for food. The hungrier the chick was the more intensely it would beg.

During the middle of my second field season I was rather puzzled to find many of the adult birds perched at the nests over the course of several days. The weather was so good I expected them to be taking advantage of the ideal conditions to catch food and not being idle at the nests. It finally dawned on me that the adults were not out collecting food because the nestlings were 'full'. The nestlings were no longer begging. They had simply eaten enough. Such long fine periods must provide a much needed breathing space for the hard working swifts!

Chapter 6
Life in the Nest

Nests are warm and safe. Adult birds return regularly to the nest during the day and nights are also spent at the nest. Once hatched nestlings are constantly present until they fledge. This means nests provide an an ideal habitat for a range of parasites eager to exploit such predicatable resources.

Louse-flies

Swifts are parasitized by a revolting but at the same time beautifully evolved insect known as the Louse-fly. These insects live off swift blood which they suck from nestlings and adults. They are nest parasites, more closely associated with the nests than with the actual birds. If swifts can be said to be superbly adapted to an aerial lifestyle, then these Louse-flies are equally as well adapted, but as parasites with a lifestyle finely tuned to the hosts.

Biologists have a strong interest in parasites. The specialized lifestyles parsites have can be used to study a range of ecological questions. Additionally there is an applied interest. By understanding how any parasite lives, we can better understand those that directly affect man.

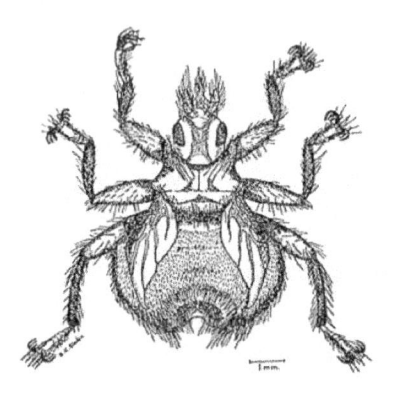

Drawing of an adult Louse-fly from David Lack's book, 1956.

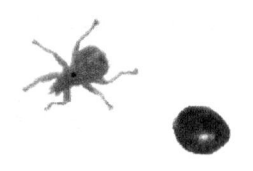

An adult and a pupa.

I quickly realised the potential Louse-flies offered me to study a range of questions. They had been little studied before, probably because few people had access to the swift nesting sites where they were found. I first became aware of the small black pupae during the winter months. These were scattered around the chambers. They resembled small black shiny balls. I was eager to see the first adult parasites themselves when they hatched! I searched nests daily and shortly before the arrival of the first swifts I found my first adult Louse-fly.

Louse-flies are superbly adapted for feeding on blood and remaining on the hosts. They are flattened in profile which makes it easy for them to crawl between the feathers to feed. The body is covered with short stiff hairs which allow them to tangle easily with swift feathers and thus remain firmly attached to the hosts. The wings have become much reduced in size. They are so closely associated with the nests and swifts that they no longer need to fly themselves. However wing stumps remain as these provide a further point allowing them to hook onto swift feathers.

The legs are equipped with sharp points aiding their attachment to swifts. Louse-flies are relatively large insects, measuring nearly a centimetre across. The mouthparts are equally well designed. There is a narrow tube to suck up blood. Through study I found that they feed roughly once every five days. Hungry Louse-flies would have meagre abdomens which were light brown in colour. After feeding the abdomens would become enlarged with blood looking like little round balls. These have a grey colour.

The Louse-flies were concentrated around the nests. This is understandable. Being carried away with the adult swifts is not a good idea. You risk falling off the swift a long way from a nest, and then you would be a long way away from your source of food. Nests are warm and safe and offer opportunities to feed regularly so it is best to stay in them..

The nestlings seemed to be their main target for feeding. It was not uncommon to see a nest covered with Louse-flies. These would quickly run all over a nest when they sensed the slightest movement. Such a nest crawling with these parasites was quite a revolting sight to behold. Especially when you had to put your hand in it to extract a nestling.

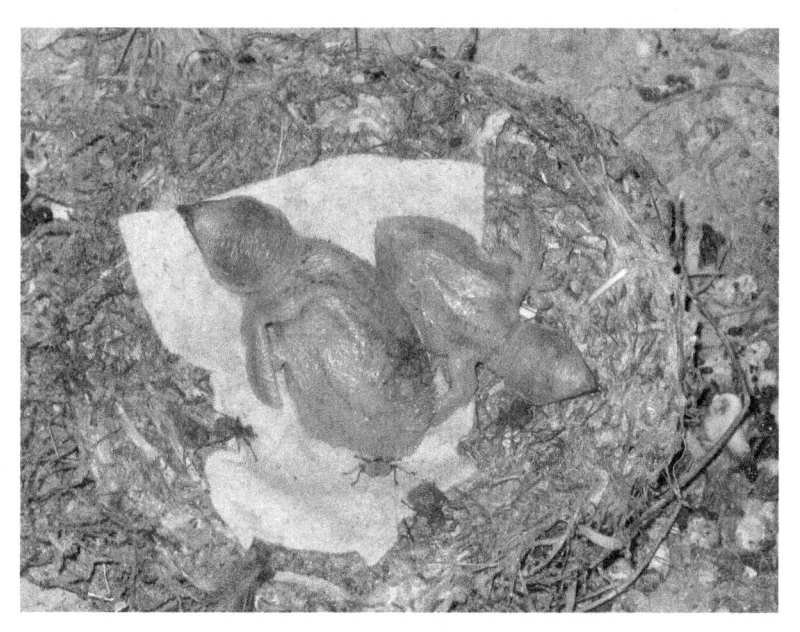

Note the Louse-flies around the nestlings.

As is typical with parasites, some nests were extremely heavily parasitized, sometimes with as many as 40 Louse-flies. Others had none. They seemed however to be easily transferred between nests. When I marked them they soon reappeared at nests some considerable distance away. It appears they were carried by adult birds in either the feet or feathers between nests.

The life-cycle of Louse-flies is closely linked to that of the host swifts. If swift breeding is greatly time constrained, then that of the Louse-flies is even more so. The Louse-flies overwintered as pupae which as mentioned were small shiny black balls that were found around the nests. They only hatched out into adult Louse-flies with the return of the swifts in the spring. A period of frenzied breeding would then occur, with nests shimmering with Louse-flies many engaged in mating. However populations soon peaked and began to drop. By the time the swifts left nearly all of the Louse-flies had completed their life-cycle and died away.

The main aim of my research was showing that these insects have a detrimental effect on nestlings. As they feed once every few days, are relatively large parasites in comparison wth the nestlings, and can occur in large numbers, they should have a strong effect on the swifts. But when I compared growth of nestlings between heavily parasitized and under parasitized nests I could find little difference. This was surprising.

I also spent much time studying the Louse-fly lifecycle in detail. Like many parasites Louse-flies are greatly aggregated, being concentrated in a few unlucky nests. I also studied how Louse-flies move between nests by putting numbered discs on adults.

But the Louse-fly is not the only pest the swift has to contend with. Another parasite which I found more revolting than the Louse-flies were the small brown lice. These would crawl over the head of the nestlings and sit on the side of their eyes apparently feeding off the fluid of the lacrimal glands.

Torpor

The nestlings relied on a fairly constant supply of food from the parents. But what should they do if this supply suddenly stopped? As we know, summer weather conditions in Europe can be most unpredictable. Periods of cold wet weather are almost predictable! Nestlings are almost guaranteed to experience days when food is short in supply. What should you do during such times? One method to survive such periods is to downregulate metabolism, decrease body temperature, and basically fall asleep. This helps save precious energy and might allow you to survive until conditions improve and a regular food supply starts again.

Scientists have speculated that swift nestlings can enter a state of torpor, with extremely lowered temperatures and lethargy. Torpor has even been seen in a number of swift species, including the White-throated needletail and White-throated swift.

During extended cold periods I often noticed that although initially nestlings were tolerant to cooling, over time they would get gradually more and more lethargic. They became less active when exposed for several hours. When held in the hand they would be cold to the touch. Many would appear to be asleep in the nest, being completely motionless. This certainly appeared to be a form of torpor.

Parental Care
The swift seems to be a model of sexual equality. Incubation is done jointly, with pairs taking it in turn to warm the eggs. When the nestlings hatch out the

adults take it in turn to shuttle to and fro with packages of food for the young. There appeared to be little difference in the amount of care between the sexes.

Some pairs were really poor parents. I guessed these were mostly young pairs who still needed some practise before being able to be really successful at breeding. For example one pair began to build a super nest with a tightly woven outer ring. However when the nestlings hatched and I could examine the nest more closely it transpired that the pair had omitted to put a bottom to the nest. The nestlings were sat on the bare cold concrete which provided no insulation whatsoever.

Another sign that this pair was inexperienced was that they left the nestlings totally unbrooded during the day. Whenever I would visit the nest the nestlings would always be stone cold and inactive. One of the pair seemed to have considerable difficulty in orientating between the nest and the hole. Although the nest was very close to the exit. I would often find this bird in corners of the chamber or hanging from the wall some distance away. Needless to say these nestlings eventually died. This pair had anyway begun incubation very late, always a bad sign, so it would have been unlikely that the nestlings could have grown and matured quickly enough anyway. Late hatched nestlings had much lower survival chances and grew slower than nestlings that had hatched earlier.

Adults varied in their behaviour towards me. Some became aggressive when they saw me at the nest. Often they would peck out vigorously when I fished for the young from under them to weigh and measure. Most however were scared and would scuttle away. Others just sat while I went about my business around them. Some were more bold and would come over and see what I was doing.

I often noticed that several parents seemed to return simultaneously or at least within a few minutes of each other. Although judging this was difficult as I could only be in one chamber at one time. Are swifts communal hunters? It appeared so to me. This would make good sense. Aerial insects are probably an ephemeral resource occuring in patchy cloud like blooms. Finding such areas is probably difficult. Following successful swifts who know where such resources are to be found would be a good strategy for a hungry swift.

An adult swift.

Sexual Equality

Studying differences between male and female swifts is difficult. This is because males and females are identical in appearance. There is no way of knowing which is which! Both male and female appeared to spend equal amounts of time on the nest incubating and providing similar amounts of food to the young.

Many bird species were traditionally assumed to be monogamous forming close pair-bonds. But the use of genetic techniques showed that extra-pair mating occurred in many species with males rearing young they had not parented.

The swift is also thought to be monogamous. As of yet no evidence of extra-pair matings has been found in swifts. However, genetic study comparing young with their parents might find some. I observed that birds were most likely to be aggressive at the nest sites. This could be males protecting their mates in the most practical manner possible. It is likely most mating occurs at the nest sites so they have an interest in keeping other males away. If any extra-pair mating does occur it is likely to happen here at the nest sites.

It could be that there are differences in how males and females provide for the young. It would pay one partner to leave the lions share of care to the other partner. However, my observations showed little apparent difference in provisioning between pairs, albeit I did not know which were male or female. The problem for any swift wishing to shirk parenting is that by reducing the amount it provides it could well be reducing the chances of survival of its own young. Swift rearing is on such a knife-edge that the risk of losing your reproductive contribution for a year might not be worth a relatively meagre gain in food.

Chapter 7
Teenager Troubles

As the chicks got older they got more difficult to work with. From a few days of age their grip strengthened and they had to be more or less prised from the nests to be weighed and measured. Care had to be taken when lifting them out so that they were not injured. Despite their strong hold on the nests the legs and feet were still tiny and feeble and I was worried lest they be easily broken.

To illustrate how strong this grip was, at some nests which were not well attached, it was possible to raise the entire nest from the ground simply by lifting the nestlng up. Often the claws would become entangled with the threads making up the nest and the nestlings had to be carefully 'unhooked' from the nest. That the nestlings should want to remain in the nest is perfectly natural. Inside the nest they were safe, while outside of it they faced a range of dangers.

Another problem I faced was that as they aged the nestlings became increasingly fidgety. They just would not stay still once you had them in the hand. This made taking accurate measurements difficult. The wings were especially hard to measure. Only rarely would nestings willingly spread them out for me. Each measurement was taken three times to try to ensure accuracy. Later I worked out an average from all three measurements.

It took a good six to seven hours to measure and weigh every nestling in every nest. Not surprising when you consider I had over 50 nestlings to measure and the range of figures I took. As well as weight I measured wing length, leg length, head length and width, and toe length. Plus I counted Louse-flies on each bird and each nest. As they grew older work began to take even longer. When the feathers began to grow it became even more difficult and even more time consuming.

In my first year, 2007, there were 38 breeding pairs at the bridge which produced 75 young. But the colony was slowly increasing in size. In 2008 there were 41 pairs at the bridge,. These produced 89 nestlings of which 38 fledged. 2009 was the best year with 45 nestlings fledging from 38 pairs. With a minimum of disturbance the colony could get even bigger.

10 to 15 days of age
As they became older the nestlings became more adventurous. They started to begin to leave the nests and wander short distances from them. They returned to the nest after a few minutes or when a parent returned with food. These short excursions could sometimes end in disaster.

One day I visited one nest to find two nestlings where the day before there had been three. A quick search in the chamber revealed the missing nestling had not got lost in a corner or gone to another nest. It appeared to have vanished. What must have happened is that it had gone wandering and fallen through a hole and drowned in the water below. Maybe it had been curious or perhaps had tried to get a head start on its siblings by getting closer to their returning parents with food.

The urge to move around and walk about varied between different nestlings and different nests. Some would remain happily sat in the nest and would rarely move away from it. In other cases chicks were much more eager and ready to stumble around.

Normally in swifts the parents stop brooding the young during the day once the nestlings reach seven to ten days of age. I had hoped the bridge would be adult free at this time allowing me to check the nests quickly and without disturbance, but this was not the case. Parents were still often present and hindered my work. To avoid disturbing birds unduly I would often have to sit motionless for long periods of time while adults fed young and left.

Nest swoppers
Nest swopping is where nestlings move between nests and try to become 'adopted' by other parents. It was an uncommon occurrence at my colony. The usually long distance between nests made it difficult for

nestlings to reach others nests. Often nests were seperated by chamber walls. Finding neighbouring nests in the darkness was difficult. It would also be a dangerous endeavour. Should a nestling leave its home nest to search for another it might become lost and disorientated within the chambers and easily fall down a hole. Or its parents might return to the nest and on finding the nest empty not return again assuming the nestling had died.

However there were some exceptions when it did occur. I would come up to a nest expecting say, to find a single small nestling, only to find the nest engulfed by a much older feathered nestling. The unfortunate native would normally be squeezed underneath the squatter. Or I might come to a nest and expect to find two nestlings and instead for there to be a third; all three of similar size. In such cases it could be extremely difficult deciding which the adoptee was and which the native nestlings were.

In all I only had several instances of nest swopping each season. When I came across it I always noted the fact and often returned the swopper to its original nest. Those that had nest swopped once were likely to reoffend. Most nestlings were quite happy to remain in their own nest, but a small minority would do it over and over again. These nestlings always seemed restless when I visited them at the nests and were always the ones to scutter away. They seemed pre-destined to nest squat. Maybe such nestlings possess genes for such delinquency.

One of the reasons swifts like to have colonial nesting sites could be that in emergencies nestlings have the possibility to find a new nest through such wandering. This could be beneficial if parents came to a sticky end. Adults which nested colonially and produced offspring with this ability might over a very long time have a slight advantage over parents whose offspring did not do this. At least if your nestlings move to a new nest you no longer have the trouble of feeding them anymore.

There may be a number of other reasons causing nestlings to nest swop. Nestlings may not like the home nest and may be seeking better pastures elsewhere. Maybe the native nest is heavily parasitized. Or maybe the parents are poor providers and the nestling wants to find better ones. Maybe there is simply too much competition from siblings at the native nest. If some of these things are true you might expect swopping to occur most frequently at heavily parasitized nests or during poor weather when the costs of sibling competition, parasitism etc. are at their greatest.

Even young swifts could climb.

However my own experience, which is only anecdotal and not based on data, is that there is no connection between these things and nest swopping. If anything nestlings seemed to me to be more active when the weather was good and it was warm. They had surplus energy at these times which they could channel into moving around. When it was cold they instead tried to save energy by not moving and staying in the nest. Likewise most nestlings fell through the holes during periods of good weather when they were most active.

Food obsession

The young chicks were food obsessed. When they heard a parent returning with food they would go mad in the nest, screaming for food and waving their open mouths around. I induced a similar reaction when I went up to the nests and tried to remove a nestling.

The parent birds returned on average once every hour or so with food. Their cheeks were always full and bulged out of the sides of their heads because of the large amount of food they were carrying. It looked as though they had extremely swollen glands or had swallowed two small balls and were keeping them in their mouths. The rate of delivery of food depended on the weather and the resulting abundance of insects.

For obvious reasons it was difficult for me to see what exactly the parent birds were feeding their young. The parents would transfer the food from the cheek pouch directly into the gape of the chicks. Luckily on some occasions I did get to have a look. Often when a parent bird flew in and became spooked at seeing me it ejected its food pellet and quickly left again. I was able to simply pick the pellet up from the floor. They often turned out to be a compact mass of mixed up insect remains.

Determining exactly what type of insects made up the pellets was impossible, at least to my inexpert eyes. They mostly seemed to be small in size and similar to the little midges you see flying around. I was surprised at how compact and firm the pellets were. The adults must really pack the insects in. Nevertheless I was easily able to prize them apart with my fingers.

Clutch size and nestling growth

In clutches of three nestlings two would be similarly sized and the third much smaller. Sometimes there might be one particularly large nestling and two smaller siblings. In clutches of two nestlings they could be either similarly sized or one could be much larger than the other. Single nestlings seemed to grow more rapidly than nestlings in clutches of two or three. This makes sense as these nestlings were receiving twice or three times as much food.

Size probably depended on when chicks hatched. When both eggs hatched simultaneously each chick received similar amounts of food. However when one chick hatched earlier than its single sibling it had time to grow in strength so that it could monopolise the food supply to the detriment of its sibling.

The chicks first to hatch were the biggest. Where there was a small weedy chick it had obviously hatched later than its siblings and because of its small size had had trouble competing for food and so remained small. Brotherly love and fair sharing does not occur in swift families. Each chick only cares about itself.

At a couple of nests where there had originally been three chicks after a few days the weaker third chick would be found dead outside the nest. In one case I found the chick over 30 cm away from the nest. I wondered whether this was a case of programmed suicide. It would be best to leave the nest if you knew your chances of survival were too small rather than let your parents waste energy on you. Thus your parents could concentrate on caring for your siblings instead.

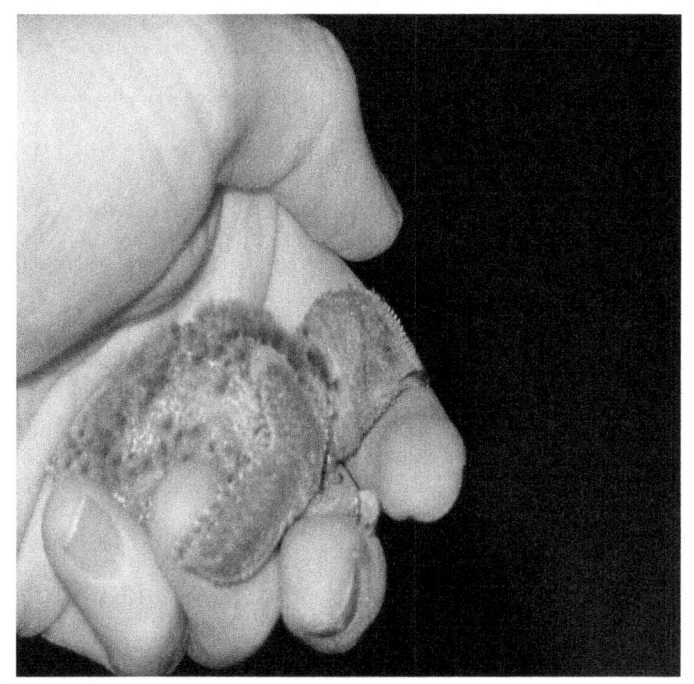

A young chick.

However, it is more likely that the older stronger siblings forcefully ejected their weaker nest mates, who could not fight to get back in again. There was indeed much jostling and pushing for position within the nests and I could well imagine weak chicks having a hard time of it.

15 to 20 days of age
From about day 15 of their lives the baby swifts took on a much more adult like appearance and seemed to mature rapidly. In form they began to resemble more and more the adults. The neck became shortened, the eyes sunken into the sides of their heads and they took on a sleeker aerodynamic appearance. The resemblance to a ball with a mouth on one side was lost.

The nestlings were now so big that they had problems fitting into the nest. This was especially the case when there were three young in a single nest. In such cases one of the young would sit in the nest and the other two would sit beside it outside the nest.

20 days onwards

Nothing looks sweeter than a 20 day old swift nestling. The feathers are almost fully grown and are soft to the touch and have a new clean appearance. The white chin patch has developed and this gives the chicks an almost cheeky like appearance. At this age the nestlings look like miniature copies of their parents but the head is rounder. They sit most of the day within the nest, and have grown to such a size that they now completely fill it. Their head sticks out at one end, while their tails stick out 180 degrees opposite.

In the last stage of their lives in the nest I noticed a slight change in behaviour in the young swifts. Normally they would be quite placid in my presence, but now they were more suspicious and would try to scutter away from me if I came too close. I also noticed that at about this time the nestlings would call outside in reply to underflying swifts.

Note the stumpy wings and tail in the nestlings.

It was strange to think that the chicks, now really sub adults, had spent their entire lives only in the dark confines of the bridge chambers. I had had enough of the claustrophobic atmosphere after a few hours, but they had spent weeks there. They had never been outside and had no experience of daylight or sunshine. The most they had experienced of the outside was when the sun shined through the drainage holes and reflected a circle of light on to the chamber roof above.

Despite never having experienced it, they were however totally dependent on the outside world for survival. Whether their parents could find enough food for them depended on the weather conditions outside. Each time the parents left they did not know when, or even if, they would ever come back.

As they became older, working with the nestlings became increasingly difficult. Stroppy adolescents. They would continually wriggle in my hands trying to get free. This meant taking accurate measurements was not easy. When they saw me approach they would begin to scuttle away. But there were pluses. The swifts were now at their most sweet and adorable. The newly developed white bibs but still fairly squashed faces gave them a very characteristic cuteness. This was a considerable change from the ugliness of the newly hatched nestlings.

When I told people where I was working, by the beautiful Bigge reservoir, they would often exclaim 'how nice'! They were thinking of the wide expanse of water, the rolling wooded hills in the background and the peacefulness of the surrounding countryside. Actually my working conditions were not as pleasant as these people assumed.

I could see nothing of the reservoir or the wooded hills from within the bridge. Instead I was working in pitch darkness with noisy traffic overhead. After a few hours of being inside the bridge you soon became conditioned to working in the dark and with the noise. It was only when you re-emerged that you realised it was a nice sunny day outside. I became immune to the constant hum of traffic. My ability at walking around in the dark and of doing things without seeing what I was doing improved considerably.

Chapter 8
Fledging

Fledging is the term used to denote nestlings that are near to gaining independence. They have just left the nest, or are ready to do so. Fledglings are ready to fly, or almost, and can pretty much live independently.

Fledging is often mistakenly considered as being the end point of growth. This is not true. The fledglings still have to develop somewhat before reaching the physical dimensions of the fully grown adults. This is apparent when you have the opportunity to compare an about to fledge nestling with its parent at the nest.

It is not so much that the fledglings have to grow much bigger to become like the adults. Rather their physical form needs to mature. The head needs to become even less rounded and more flattened in profile. The body needs to become thinner, sleeker and stronger looking. In colour the nestlings have to darken somewhat from dark brown to the adult pure black.

Two nestlings sat together.

Even at 20 days of age the nestlings are rather podgy with large rounded bulging bellies. They weigh roughly between 35 and 45 grams, which is almost the same as adult weight. However before leaving the nest they first put on another 10 grams or so, reaching say 50 grams, before 'dieting' in the last few days in the nest and leaving at the normal adult weight at around 35 days of age.

Such weight regression is characteristic of nestling aerial insectivores or seabirds. These species have to be fully capable of perfect flight on leaving the nest. During the last few days in the nest the excess fat supply which would hinder flight is lost, and the flight muscles are toned and strengthened. When swifts leave the nest they are lean, mean flying machines.

Most species of bird do not exhibit such weight regression at the end of the nesting period. This is probably because although they leave the nest they do not become fully independent but continue to rely on help for several further days from the parents. Swifts on the other hand receive no further feeding or help once they fledge. They become fully independent and must be fully capable of flight at once.

The flight capabilities of the young swifts seemed to develop in some cases a good few days prior to them leaving the nest. At ages 28 to 30 days of age nestlings dropped (carefully) about a half a metre on to my jumper were not able to sustain any kind of powered flight or gliding. Although they flapped the wings vigorously they would simply fall vertically, with the flapping only helping to break their fall.

In just a few days, by ages 30 to 32 days, the capability for powered flight seemed to develop considerably. Instead of falling vertically downwards dropped swifts began to be able to fall more in an arc. The vigorous flapping was now starting to have some effect and allowing the nestlings some forward propulsion. Although proper self-sustained flight was still obviously some time off. Now when dropped nestlings landed some two or three metres distant, being able to 'glide' this distance while flapping. The wings had hardly grown in size, but nestlings had lost over a gram in weight. This shows the importance of weight and wing loading as opposed simply to wing size. The wings at this stage measure between 110 and 140 mm in length. Those of swifts which fledge are in the region of 160 mm.

Some authors report seeing young swifts at this age doing 'press-ups'. Apparently the swifts spread their wings out wide and then repeatedly raise and lower their bodies. This phenomenon has led to speculation that this is a form of exercise helping to tone the flight muscles and prepare them for flight. However, such training is not a prerequisite for successful flight. Nestlings which are raised in nest sites which are too narrow to allow such exercises are still capable of fledging and flying perfectly.

I was never lucky enough to observe this behaviour. However I do wonder if people who have seen it are mistaking it for something else. When young swifts beg they often spread their wings out and flap them forcefully against the ground. I can't help wondering if this begging has been mistaken for something else. My presence at the nest, or even a slight noise similar to adults returning was enough to evoke this begging response.

Near the end of the nesting period swifts were happy to just sit in the nest.

I found one of the most distinctive aspects of the physical change which young swifts undergo in their final stage at the nest to be the change which occurs to the shape of their heads. At 25 to 30 days of age the head simply appears to widen. The eyes are very noticeable at this age. As is the prominent white chin patch. Where the head attaches to the body is very clear to see. However by 32 days of age the head has changed in shape and attachment to the body considerably. Now it appears much narrower and slender. The eyes have become sunken into the head in the typical adult 'augenniche'. Also differentiating between head and body becomes difficult. The join between the two becomes indistinguishable. The nestling swift now has the 'torpedo' bauplan of the adults.

Some fledglings would reach fledging size and then fledge almost immediately. I might notice they were approaching the appropriate size, and then find them gone the very next day. In some cases I might not realise the nestlings were ready to go, until I found an empty nest and figured out what had happened. In other cases the nestlings would seem to linger! They would reach the correct fledging size and I would expect them to be gone a day or two later. Instead each day they would remain in the bridge.

A direct gaze.

In the second year I had half a dozen nestlings which I expected to be gone or be going within a day or two- but they stayed three or four days longer than I expected. I was a bit surprised because the weather was good and I thought they would want to leave to take advantage of it. But a day or two later a strong 'low' swept over us and brought heavy rain for much of a day. After this weather had passed the nestlings seemed to finally leave. Could they sense the approaching bad weather and delay their leaving accordingly? I would not be surprised if they were intelligent enough to be able to do this.

Such late leavers showed me that the nestlings continued to grow at a similar rate after fledging as they did just before, at least initially. Most fledglings would leave once their wings had reached a size of 160 mm and their primaries a length of 125 mm. But these lingerer's would continue to grow in the following days, often gaining wings of 170 mm and primaries of 135 mm in length.

I already knew that nestlings were pretty much capable of flight once the wings had reached a length of 150 mm (depending on their weight though). These things told me that nestlings do not necessarily fledge as soon as flight is possible; nor when the wings have slowed in their rate of growth. During the last few days in the nest I would observe the young swifts increasingly spreading their wings out; as though they were practising flexing their wings and seeing what they could do.

Although the nestlings are capable of some form of powered flight maybe this is energetically expensive at this point. Better to wait slightly longer until the wings are a little longer and flight is more efficient. I would not be surprised if nestlings fledge when the costs of flight can be equalled by the amount of food they can catch. However testing such an idea would be impossible.

Close to fledging the wings would become very long and stick out behind the nestling.

Another reason delaying fledging could be that the nestlings are somewhat lazy! As we all know being independent and free is exhilarating, but it is not easy. You have to pay your own keep. In the nest the young have food brought to them on a regular basis and do not spend energy on flying or hunting. Once airborne they must remain airborne; a possibly costly affair. So just like stay at home children who stop with their parents until their late 20's, the nestlings could be making the best of an easy life.

Of course this is of no benefit to the parents who have the considerable task of providing for their lazy offspring. One might expect parents to cease feeing at some point. However, my observations showed that this only rarely occurred. The parents would continue feeding until the nestlings had fledged and even beyond. I would often find food pellets in the nest left by the foraging adult who had returned to the nest to find it empty. They would deposit the pellet of food, fly off, and not return.

I never knew when fledging would exactly take place. I learnt to watch for the weight regression over two or three days when chicks were over 30 days old. Once this occurred I considered that the nestlings had fledged if I went to the nest only to find it empty.

My first summer had been long and hard. But now, it was the middle of July and only a single nestling stubbornly remained. It was not the only reason I was continuing to visit the bridge every day; I also had various things to tidy up and finish off. Today was hot and sultry, and the bridge appeared totally empty of swifts. By this time many had left; the adults not returning once the nestlings had fledged. I had not seen an adult for some days now. There was only a single nestling left.

I diligently visited the single nestling daily taking measurements and weighing it. Weight regression had taken place and the nestling was to all intents and purposes ready to leave. As I approached today it was sat on the edge of a hole, as nestlings often did, looking out at the sunny day outside. What was it thinking? Are nestlings scared of taking the plunge? I took my measurements, but the nestling did not undergo the normal bout of fidgets that accompany handling. It was quite happy and remained composed as I stretched its wings out and measured its head.

When I replaced the chick in the nest it ruffled its feathers in seemingly irritation at being held as they normally do. What did the future have in store for this swift? Would it see the Sahara, African skies and return in a few years to raise its own young? Would it still be here tomorrow? I returned to my other tasks around the bridge, finally leaving an hour or so later.

After being cooped up inside for a number of hours, I often paused at the banks of the reservoir, enjoying the fresh air and light, and looking at the bridge from the outside. Today being particularly nice meant I particularly enjoyed this pause. I gazed up to the bridge, my eyes scanning for the chamber where the nestling remained. Why was it so reluctant to leave when it was so nice outside in the open where it could swoop and play in the air? I sat for some time gazing upwards, then suddenly I thought I saw something, a dark small bundle, come from the bridge. Was it the nestling? Had it decided to make the jump? Or was I simply deceiving myself? On my next visit to the bridge the day after, the nestling had gone and I was alone in an empty bridge.

GOODBYE SWIFTS!